Silkworm Moths

by Dina Drits

Lerner Publications Company • Minneapolis

For Brian, Beetle, and Bailey
—DD

The photographs in this book are reproduced with permission from: Photo Researchers, Inc.: (© Harry Rogers) front cover, pp. 14, 17, (© Danny Brass) p. 9 (left), (© Ray Coleman) p. 9 (right), (© E. R. Degginger) pp. 24, 32, (© David M. Schleser) p. 25, (© Tom McHugh) p. 26, (© Pascal Goetgheluck/Science Photo Library) p. 33, (© Jeff Greenberg) p. 36; © Bruce Coleman, Inc.: (© Wardene Weisser) pp. 4, 11, 18, 35, (© E. R. Degginger) pp. 6, 10, 46-47, (© George D. Dodge) p. 8, (© Rex A. Butcher) p. 43; © Robert & Linda Mitchell, pp. 5, 20, 27, 29, 31, 37, 38, 39, 40 (both), 41, 42; © Dwight Kuhn, pp. 7, 12, 13 (both), 19, 21, 22, 28, 34.

Lerner Publications Company
A division of Lerner Publishing Group
241 First Avenue North
Minneapolis, MN 55401 U.S.A.

Website address: www.lernerbooks.com

Library of Congress Cataloging-in-Publication Data

Drits, Dina.
    Silkworm moths / by Dina Drits.
        p.   cm. — (Early bird nature books)
    ISBN 0-8225-0069-8 (lib. bdg. : alk. paper)
        1. Silkworms—Juvenile literature. [1. Silkworms. 2. Moths]
    I. Title. II. Series.
SF542.5 .D75 2002
638'.2—dc21   2001001358

Manufactured in the United States of America
1 2 3 4 5 6 – JR – 07 06 05 04 03 02

Contents

Be a Word Detective

Can you find these words as you read about silkworm moths? Be a detective and try to figure out what they mean. You can turn to the glossary on page 46 for help.

abdomen

antennas

cocoon

compound eyes

domesticated

exoskeleton

larva

mandibles

molt

prolegs

pupa

simple eyes

spinneret

spiracles

thorax

Chapter 1

An adult silkworm moth and a young silkworm rest on a tree. Silkworms make something that people use. What is it?

This Moth Is Not a Worm!

Silkworm moths are amazing creatures. They make silk when they are young. People make the silk into beautiful cloth.

Young silkworm moths are called silkworms. But silkworms are not worms. Silkworms are caterpillars. The caterpillars turn into silkworm moths when they are adults.

Moths start out as caterpillars. This caterpillar will become a silkworm moth.

An ant is an insect. There are about 1,000,000 species of insects in the world.

Silkworm moths are insects. All insects are the same in certain ways. They all have three main body parts. Insects have six legs, too. And most insects have wings.

8

Silkworm moths are just one species, or kind, of moth. There are about 100,000 species of moths in the world. Other moths are gypsy moths, Isabella tiger moths, and peach moths.

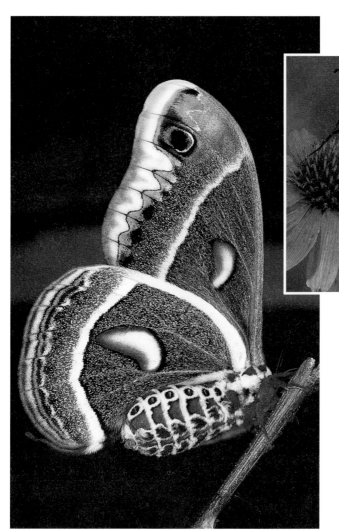

A cecropia moth (left) *and a monarch butterfly* (above) *look similar. Moths are closely related to butterflies. They both come from the group of insects called Lepidoptera.*

Moths live all over the world. They can live in hot jungles and cold icecaps. They live almost everywhere on land. Many species of silkworm moths live on the continent of Asia.

This wild silkworm moth lives in Asia. Silk from a wild silkworm is usually called tussah silk.

One species of silkworm moth is domesticated. It makes the most beautiful silk. The scientific name of this species is Bombyx mori.

Most silkworm moths live in the wild. But people raise some silkworms on farms. These silkworms are domesticated (doh-MEHS-tih-kay-tehd). Domesticated animals are animals that have been tamed by people.

Chapter 2

A silkworm moth has three main body parts. What are they?

Moth Body Parts

A silkworm moth's body is made up of three main parts. It has a head, a thorax, and an abdomen (AB-duh-muhn).

A silkworm moth has two antennas (an-TEH-nuhz) on its head. The antennas are thick and feathery. The moth can smell and touch with its antennas.

A silkworm moth has two compound eyes on its head. A compound eye is not like one of our eyes. Having a compound eye is like having many small eyes. Moths do not see very clearly with their compound eyes. They use their eyes to see movement.

A silkworm moth has two compound eyes (above). *A compound eye* (inset) *is made up of many parts that fit together.*

The wings of domesticated silkworm moths are small and weak. These moths cannot fly. They just flutter their wings and hop around.

Behind a silkworm moth's head is its thorax. The thorax is the middle part of the silkworm moth's body. A silkworm moth has six legs on its thorax.

Silkworm moths have two pairs of wings. They are attached to the thorax. Wild silkworm moths use their wings to fly. But domesticated silkworm moths cannot fly.

A silkworm moth's abdomen is behind its thorax. The silkworm moth has tiny holes called spiracles (SPIHR-ih-kuhlz) on the sides of its abdomen. A silkworm moth breathes oxygen through the spiracles. Oxygen is a gas in the air. Living creatures need oxygen to live.

How a
Silkworm Moth
Breathes

oxygen
in

carbon
dioxide
out

spiracle

A silkworm moth breathes in and out through the spiracles along the sides of its abdomen.

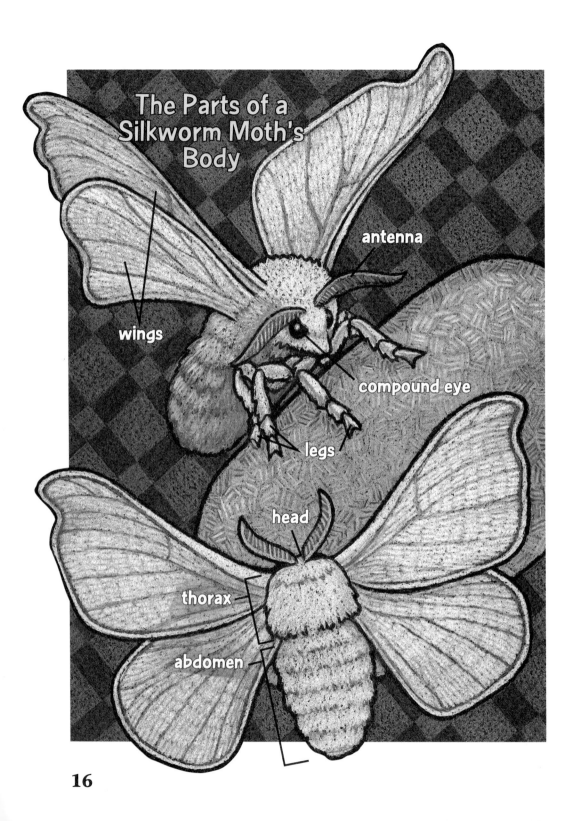

The Parts of a
Silkworm Moth's
Body

antenna

wings

compound eye

legs

head

thorax

abdomen

16

Silkworm moths do not have skin like people do. Instead, they have a skeleton on the outside of their body. The outside skeleton is called an exoskeleton (ehk-soh-SKEH-luh-tuhn).

A silkworm moth's exoskeleton and wings are covered with scales. Scales are small pieces of hard skin.

A silkworm moth is covered with tiny scales. The scales make the moth look furry.

Chapter 3

A female silkworm moth has just laid her eggs. How many eggs does a silkworm moth usually lay?

Baby Silkworms

Female silkworm moths usually lay about 500 eggs in their lifetime. The eggs are yellow and very small. One egg is about the same size as the head of a pin.

Inside each egg is a tiny larva. A larva is a silkworm caterpillar. When an egg is ready to hatch, the larva bites through the shell. Then it pushes itself out of the egg.

Silkworm eggs are sticky on the outside. They stick to whatever they touch.

A newly hatched silkworm larva is tiny.
It is smaller than a grain of rice. At first, the
larva's body is brown. But the larva turns white
soon after it has hatched.

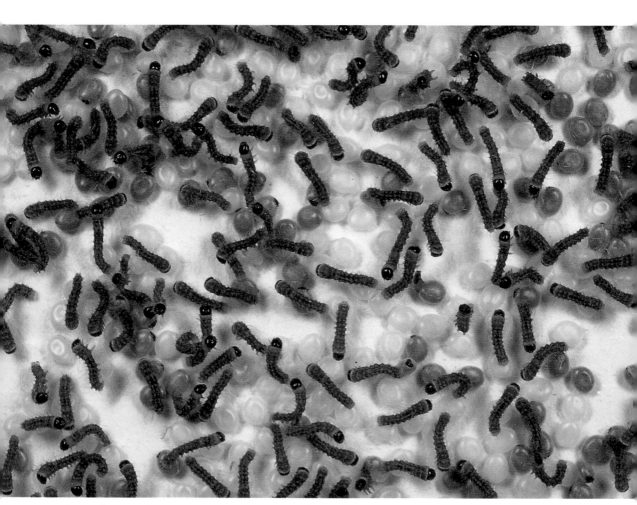

Tiny larvas have just hatched out of their eggs. Their first meal will probably be their eggshell.

This silkworm uses its prolegs to stand on a leaf. A silkworm's prolegs are curved at the end.

On the larva's thorax are six tiny legs. The larva uses these legs for climbing and for holding on to things.

The larva has 10 prolegs on its abdomen. Prolegs look like thick, stumpy legs. The larva does not use them to climb. The prolegs hold up the back part of the larva's body.

A larva has 12 simple eyes on its head. Simple eyes are much smaller than compound eyes. The larva can only use its simple eyes to see light or darkness.

Silkworms breathe through their spiracles, just like adult silkworm moths do. The spiracles are the small dots along the side of a larva's body.

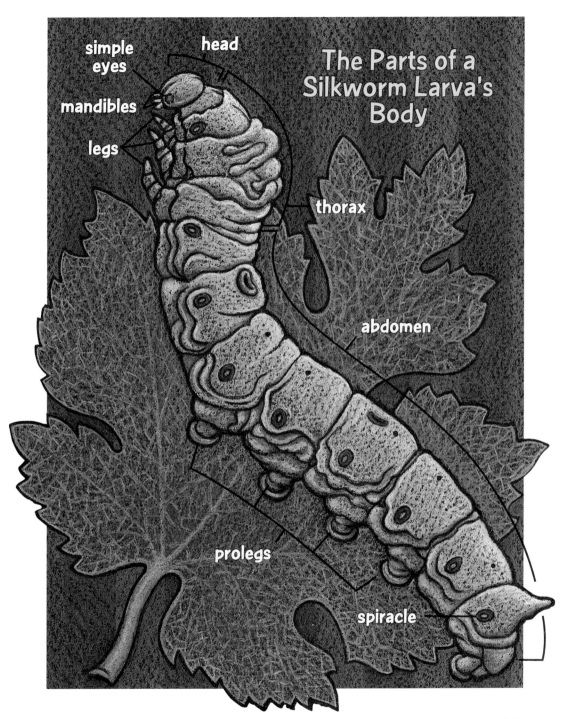

The Parts of a
Silkworm Larva's
Body

simple eyes

head

mandibles

legs

thorax

abdomen

prolegs

spiracle

A silkworm larva has strong mandibles (MAN-duh-buhlz) on its head. Mandibles are jaws. The mandibles cut and tear food. The larva has tiny teeth on its mandibles to chew the food.

This silkworm is using its mandibles to munch on a leaf.

These larvas are doing what they do best. They are eating mulberry leaves.

The silkworm larva spends most of its time eating. It eats leaves. The only leaves it eats are leaves from mulberry trees. It eats the leaves almost without stopping.

As the larva eats, it grows. Soon the larva gets too big for its exoskeleton. So the larva must molt. When the larva molts, it sheds its old exoskeleton.

A larva raises its head and stops moving when it is ready to molt.

When the larva is ready to molt, it stops eating. It raises its head into the air and stops moving. It stays very still for a day or more. Then its exoskeleton splits open. The larva slowly wriggles its body out of the old exoskeleton. The larva is already covered with a new, larger exoskeleton. The hungry larva usually eats its old exoskeleton. Then it eats more mulberry leaves.

The larva molts three more times. About six days after the fourth molt, the larva stops eating. It will not eat again. The larva is ready to spin silk.

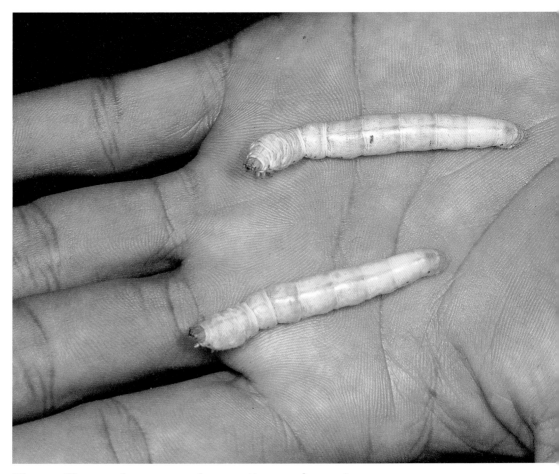

These silkworm larvas are done eating and growing. They are about 10,000 times heavier than when they hatched from their eggs.

Chapter 4

Wild silkworms often attach their cocoons to tree branches. What is a cocoon?

Making Silk

A silkworm larva spins silk to make its cocoon (kuh-KOON). A cocoon is a safe shelter.

First the larva must find a good place to attach its cocoon. Wild silkworms usually spin their cocoons on tree branches. Domesticated silkworms spin their cocoons in cardboard crates.

A silkworm larva spins a cocoon from a liquid it makes inside its body. The liquid is thick and sticky. The silkworm pushes the liquid through a tube called a spinneret. The spinneret is near the silkworm's mandibles. When the sticky liquid comes out from the spinneret, it touches the air. When it touches the air, the liquid turns into a thread of silk.

Most domesticated silkworms spin their cocoons in crates.

How a Silkworm Makes Silk

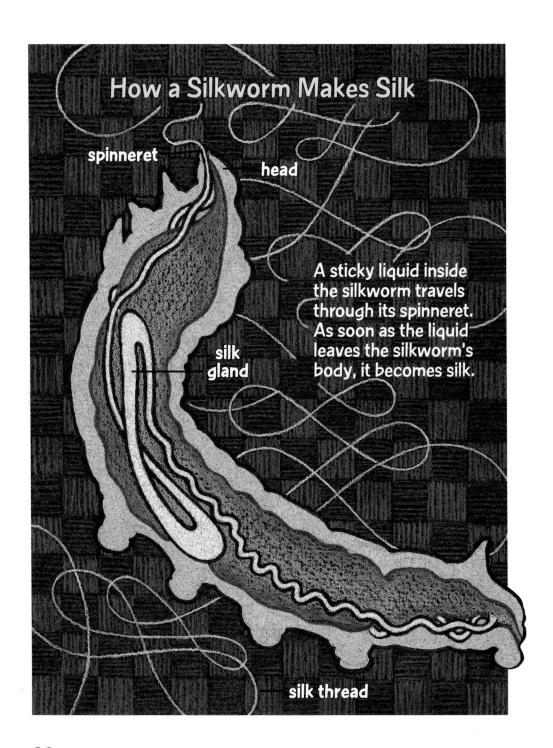

spinneret

head

silk gland

A sticky liquid inside the silkworm travels through its spinneret. As soon as the liquid leaves the silkworm's body, it becomes silk.

silk thread

A silkworm larva spins one long thread of silk to make a cocoon. First the larva raises its head in the air. Then it waves its head back and forth. The larva spins the silk thread around itself. It spins for two to four days without stopping. Finally the cocoon is finished.

These silkworms are busy making their cocoons. A cocoon is made up of a single silk thread. The thread can be up to 1 mile long.

A silkworm pupa is a different color and shape than a silkworm larva.

Inside its cocoon, the larva molts one last time. Now the larva has turned into a pupa (PYOO-puh). At first, the pupa is soft and yellow. Then its exoskeleton turns into a hard, brown covering. The pupa barely moves.

Something special is happening inside the cocoon. The pupa is growing wings, long legs, compound eyes, and antennas. It is changing into a silkworm moth.

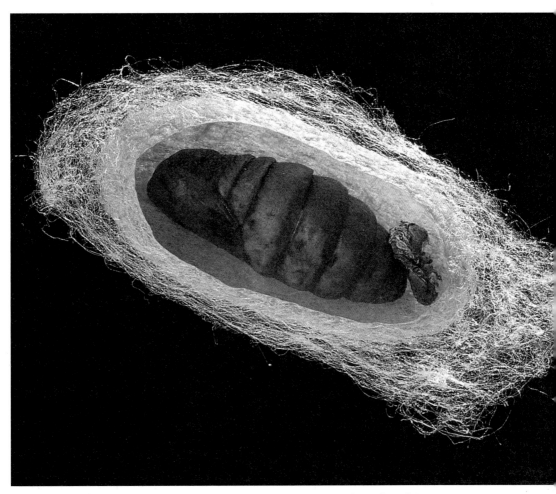

A cocoon has been opened to show the pupa inside. The skin from this pupa's last molt is to the right of its body.

This silkworm moth has just climbed out of its cocoon. Its wings are damp at first. The moth will rest until its wings are dry.

After about three weeks, the pupa is an adult moth. It is ready to leave the cocoon. The hard, brown exoskeleton splits open. But the moth still needs to get out of the cocoon. The moth spits out a strong liquid. The liquid makes a hole in the cocoon. The adult moth pushes itself out of its cocoon.

Adult silkworm moths live for only a few days or weeks. Female silkworm moths live just long enough to lay eggs that will grow into new silkworms.

Silkworm moths live only long enough to lay eggs. They do not even eat. So they do not have a mouth.

Chapter 5

Silk cloth is often made in China. Who might have discovered silk?

Silkworms and People

Many years ago, people did not know how to use a silkworm's silk. An old story says that a Chinese princess discovered silk almost 5,000 years ago. The story says that the princess was watching silkworms in her garden. They were eating mulberry leaves.

Later, the princess saw silkworms spinning silk cocoons. One cocoon fell into hot water. The cocoon came apart into one long thread. The princess dropped other cocoons into hot water. They all turned into long threads of silk. The princess realized she could weave the silk threads into beautiful robes.

Many thin threads of silk are made into thicker pieces of thread. The thread is often dyed a bright color.

After people learned how to make silk, they began to raise silkworms on farms. Silkworm farms are found in China, Japan, and other countries in Asia.

Many mulberry trees grow on a silkworm farm. The leaves on the trees will feed the silkworm larvas.

Silkworm farmers must give the larvas plenty of food to eat. They put the larvas on long tables covered with mulberry leaves. The larvas do not move much. They just eat.

A worker makes sure that hundreds of silkworm larvas have enough to eat.

Soon the larvas are ready to spin silk. Farmers have special crates where the silkworms spin their cocoons. Farmers do not let pupas leave their cocoons. The farmers kill the pupas. If they don't do this, the pupas will destroy the cocoons when they leave.

Workers place silkworm larvas in large frames (above). *The workers come back later to remove the finished cocoons* (inset).

A worker uses chopsticks to stir the cocoons in hot water. Hot water makes it easier to unwind the cocoon.

Farmers put the cocoons in tubs of hot water. Then the silk threads no longer stay together. The farmers can unwind the cocoons. Silkworm farmers sell their silk to other people.

People use silk in many different ways. They make clothing, sheets, and curtains out of silk. Silk can also be used for thread and for fishing lines. It has even been used to make parachutes!

A woman uses a loom to weave silk into beautiful cloth.

This girl is wearing a Japanese dress called a kimono. Kimonos are just one kind of clothing made of silk.

Maybe you will see something made of silk. If you do, think of the busy silkworm spinning its cocoon.

On Sharing a Book

As you know, adults greatly influence a child's attitude toward reading. When a child sees you read, or when you share a book with a child, you're sending a message that reading is important. Show the child that reading a book together is important to you. Find a comfortable, quiet place. Turn off the television and limit other distractions, such as telephone calls.

Be prepared to start slowly. Take turns reading parts of this book. Stop and talk about what you're reading. Talk about the photographs. You may find that much of the shared time is spent discussing just a few pages. This discussion time is valuable for both of you, so don't move through the book too quickly. If the child begins to lose interest, stop reading. Continue sharing the book at another time. When you do pick up the book again, be sure to revisit the parts you have already read. Most importantly, enjoy the book!

Be a Vocabulary Detective

You will find a word list on page 5. Words selected for this list are important to the understanding of the topic of this book. Encourage the child to be a word detective and search for the words as you read the book together. Talk about what the words mean and how they are used in the sentence. Do any of these words have more than one meaning? You will find these words defined in a glossary on page 46.

What about Questions?

Use questions to make sure the child understands the information in this book. Here are some suggestions:

> What did this paragraph tell us? What does this picture show? What do you think we'll learn about next? Is a silkworm really a worm? How many main body parts does a silkworm moth have? How many eggs does a female silkworm moth lay? How is a silkworm larva different from an adult silkworm moth? What do larvas eat? Why does a larva spin silk? How do humans use silk? What is your favorite part of the book? Why?

If the child has questions, don't hesitate to respond with questions of your own such as: What do *you* think? Why? What is it that you don't know? If the child can't remember certain facts, turn to the index.

Introducing the Index

The index is an important learning tool. It helps readers get information quickly without searching throughout the whole book. Turn to the index on page 47. Choose an entry, such as *eating,* and ask the child to use the index to find out what a silkworm larva eats. Repeat this exercise with as many entries as you like. Ask the child to point out the differences between an index and a glossary. (The index helps readers find information quickly, while the glossary tells readers what words mean.)

Where in the World?

Many plants and animals found in the Early Bird Nature Books series live in parts of the world other than the United States. Encourage the child to find the places mentioned in this book on a world map or globe. Take time to talk about climate, terrain, and how you might live in such places.

All the World in Metric!

Although our monetary system is in metric units (based on multiples of 10), the United States is one of the few countries in the world that does not use the metric system of measurement. Here are some conversion activities you and the child can do using a calculator:

WHEN YOU KNOW:	MULTIPLY BY:	TO FIND:
feet	0.3048	meters
inches	2.54	centimeters
gallons	3.787	liters
pounds	0.454	kilograms

Activities

Make your own silkworm moth book. Write down some interesting facts you learned from this and other books. Draw pictures to go with the words.

Pretend you are a silkworm larva spinning your cocoon. Take a sheet or blanket and slowly wrap yourself up inside it. What does it feel like? What will happen to you inside the cocoon?

Look through the clothing in your closet or at a department store. Which items are made of silk? You can find out by reading the label inside each piece of clothing. How does the silk feel?

Glossary

abdomen (AB-duh-muhn): the back part of a silkworm moth's body

antennas (an-TEH-nuhz): the feelers on a silkworm moth's head

cocoon (kuh-KOON): a safe shelter

compound eyes: eyes made up of many parts that fit together

domesticated (doh-MEHS-tih-kay-tehd): tamed by people

exoskeleton (ehk-soh-SKEH-leh-tuhn): the hard outer skin of a silkworm moth

larva: a silkworm moth in the second stage of its growth

mandibles (MAN-duh-buhlz): a silkworm's jaws

molt: to get rid of old skin

prolegs: false legs that hold up the back part of a silkworm moth's body

pupa (PYOO-puh): a silkworm moth in the third stage of its growth

simple eyes: eyes that can see only light and dark

spinneret: the tube inside a silkworm that silk comes out of

spiracles (SPIHR-ih-kuhlz): breathing holes on a silkworm moth's body

thorax: the middle part of a silkworm moth's body

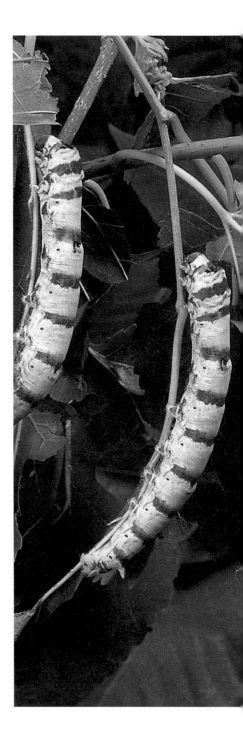

Index

Pages listed in **bold** type refer to photographs.

About the Author

Dina Drits has always loved reading and writing about animals. Growing up, she had pet mice, birds, and cats and wrote numerous stories about their adventures. After graduating from college, Dina worked as a children's book editor, where she read and edited many books about animals. She currently lives in Minneapolis, Minnesota, where she is pursuing a graduate degree in biological anthropology at the University of Minnesota.

The Early Bird Nature Books Series

African Elephants
Alligators
Ants
Apple Trees
Bobcats
Brown Bears
Cats
Cockroaches
Cougars
Crayfish
Dandelions
Dolphins
Fireflies
Giant Pandas
Giant Sequoia Trees
Horses

Herons
Jellyfish
Manatees
Moose
Mountain Goats
Mountain Gorillas
Octopuses
Ostriches
Peacocks
Penguins
Polar Bears
Popcorn Plants
Prairie Dogs
Rats
Red-Eyed Tree Frogs
Saguaro Cactus

Sandhill Cranes
Scorpions
Sea Lions
Sea Turtles
Silkworm Moths
Slugs
Swans
Tarantulas
Tigers
Venus Flytraps
Vultures
Walruses
Whales
Wild Turkeys
Zebras